muddy boots

An imprint of Globe Pequot, the trade division of
The Rowman & Littlefield Publishing Group, Inc.
4501 Forbes Blvd., Ste. 200
Lanham, MD 20706
www.rowman.com

MuddyBootsBooks.com

Distributed by NATIONAL BOOK NETWORK

Cover and interior design by Amanda Wilson

British Library Cataloguing in Publication Information available

Library of Congress Cataloging-in-Publication Data available

ISBN 978-1-4930-6568-4 (cloth)
ISBN 978-1-4930-7107-4 (electronic)

Printed in Mumbai, India
August 2022

For Mom, Queen of the Rockhounds.
And to Mike, for keeping my rock bucket full.

A behemoth thank you to my dear friends
and geologists, Marcia McConnell and
Bryan James, who tirelessly kept it real as
I attempted to reverse-engineer a rock.
—clvc

For A.O., with thanks.
—C A-F

♥ ~272 million years before she found it . . .

It began *beneath* the bottom of a *forgotten sea,*
as part of something

SPECTACULAR.

♥ ~200 million years before she ate breakfast . . .

It huddled inside a behemoth
HUNK OF SANDSTONE
held together by the mama rock
beneath the bottom of another sea.
But all that *would not last forever.*

♥ ~75 million years before she helped make snacks . . .

UP.

SPLINTERING!

UP.

FRACTURING!

Land masses
heaved UP.

FISSURING!

And it eventually split off
from the hunk of sandstone,
hurtling *farther*
and *farther* away.
No longer beneath
the bottom of BYGONE SEAS.

♥ ~66 million years before she put on her hiking boots . . .

With the impact of a great comet, life suddenly STOPPED all around it.

Wind *whoooooshhhed* past it.

Waves SLOOOP–SLOOP–SLOOPED over it.

Ice *crick-crackled,* entrapping it.

Until, the sun *warmed* the earth, again.

Rain PITTER-PATTERED, again.

And life *blossomed* once more.

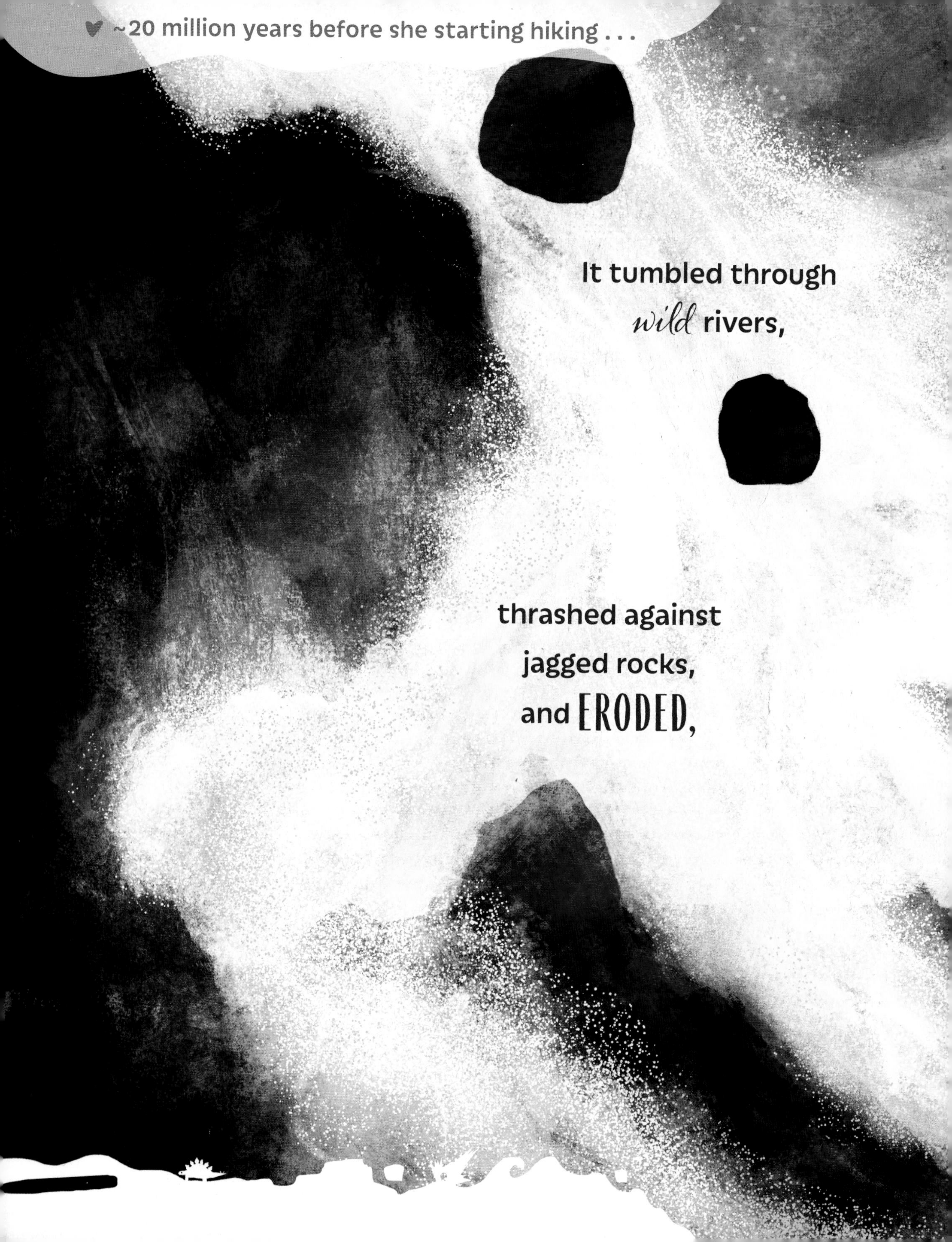

~20 million years before she starting hiking . . .

It tumbled through
wild rivers,

thrashed against
jagged rocks,
and ERODED,

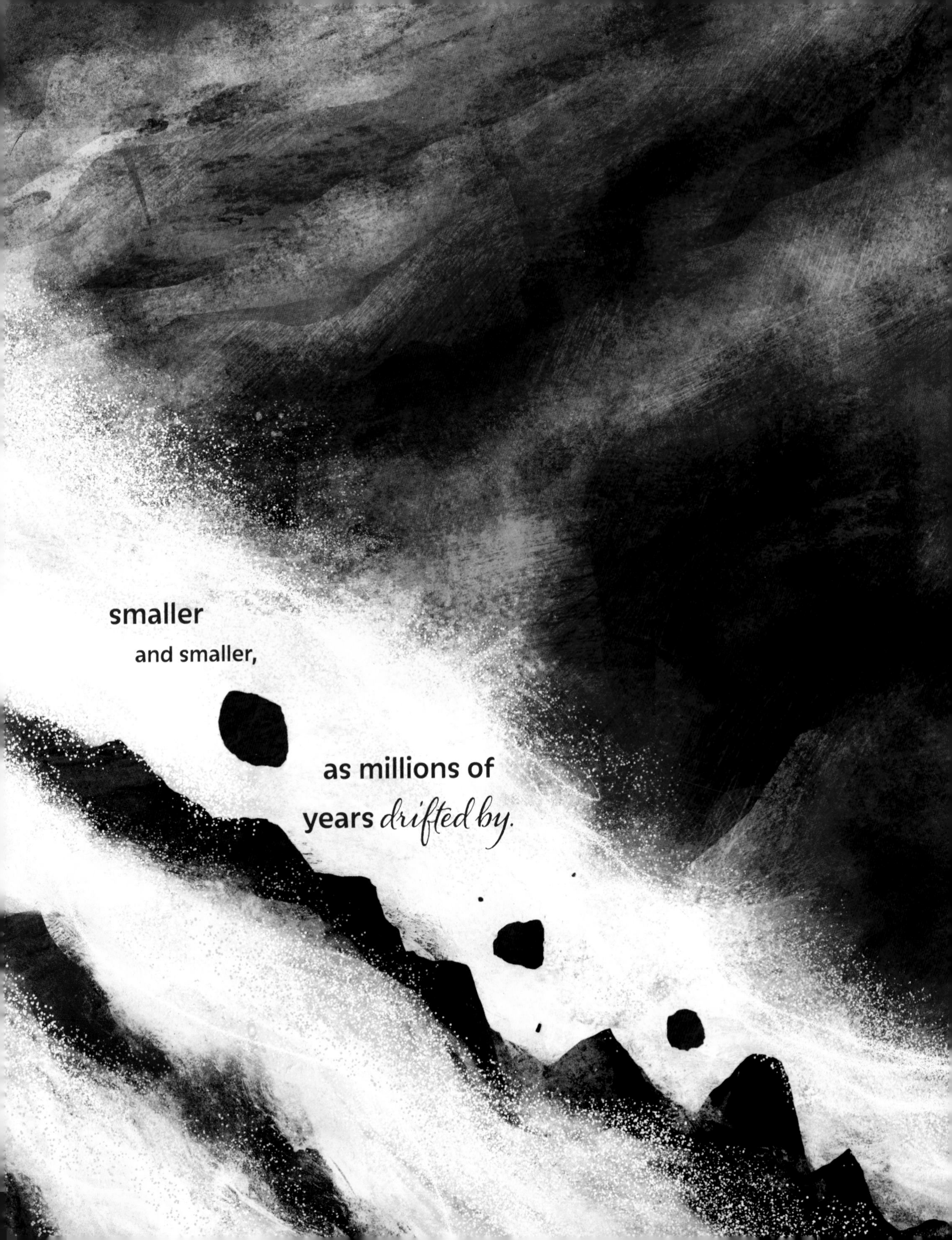

smaller

and smaller,

as millions of
years *drifted by*.

♥ ~2.6 million years before she needed a sip of water . . .
It endured
another
ICE AGE.

Wind

WHOOOOSHED.

Snow

SLOOOOOSHED.

Ice

CRICK-

CRACKLED.

~11,000 years before she saw a deer . . .
Meteor showers
SHATTERED,
launching it into
a deep ravine.

There, it rested
while everything changed
around it.
Until . . .

~10,000 years before she heard the birds singing . . .

Lightning bolts sent
it plummeting
down
down
down

toward a MIGHTY RIVER below.

~1,000 years before the wind whooooshed through her hair . . .

Naturally, the mighty river changed course.

And bunchgrass grew.

It was *pawed* and *poked* by black bears;

PRODDED and PROBED by bull bison.

Curious noses

nuzzled and *nudged* it along,

as the sun set on yet another day.

Flash floods *gushed* by,
burying it in mud.

But, it found its way to the surface again,
gully washer after gully washer.

♥ ~200 years before she hopped across a stream . . .

It was tooled by time on the shores of a grandfather river.

Smoothed.

Buffed.

ROUNDED.

♥ ~75 years before she spotted it . . .

The ever-changing terrain coaxed the ancient rock closer and closer to . . .

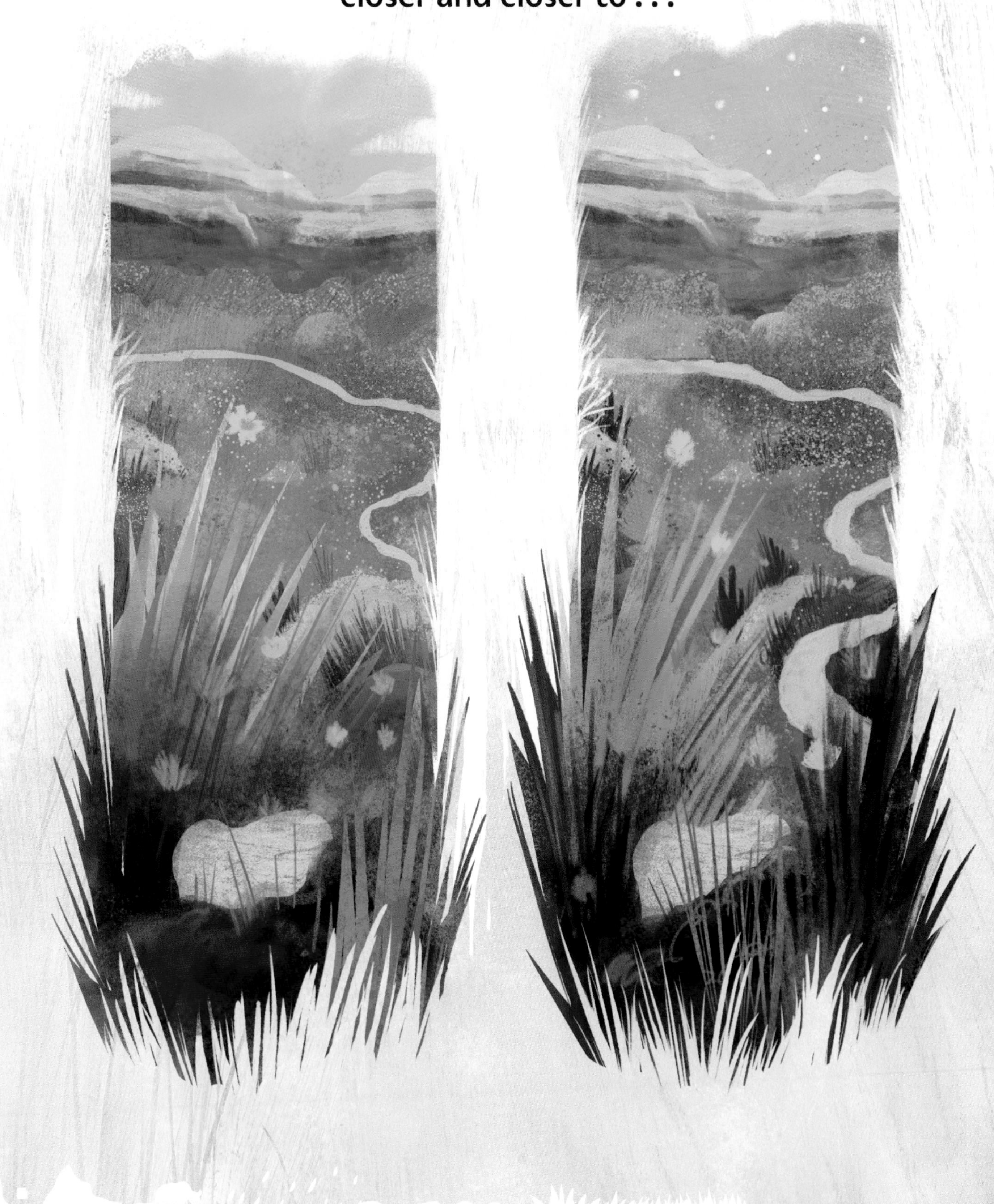

Today . . . THIS DAY!

When *it*—this heart-shaped rock—
catches the eye of a little girl searching . . .

for SOMETHING SPECTACULAR.